HARCOURT SCHOOL PUBLISHERS

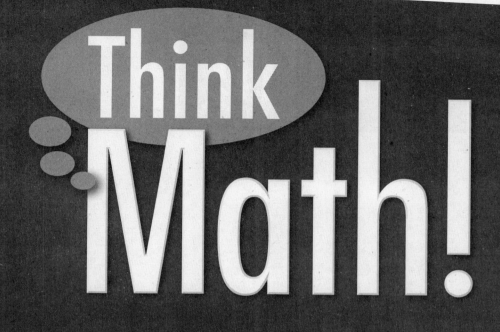

Spiral Review Book

 Developed by Education Development Center, Inc. through National Science Foundation

Grant No. ESI-0099093

SCHOOL PUBLISHERS

Visit *The Learning Site!*
www.harcourtschool.com/thinkmath

HARCOURT SCHOOL PUBLISHERS

Think Math!

Printed in the United States of America

ISBN 13: 978-0-15-342489-2

ISBN 10: 0-15-342489-3

1 2 3 4 5 6 7 8 9 10 170 16 15 14 13 12 11 10 09 08 07

This program was funded in part through the National Science Foundation under Grant No. ESI-0099093. Any opinions, findings, and conclusions or recommendations expressed in this program are those of the authors and do not necessarily reflect the views of the National Science Foundation.

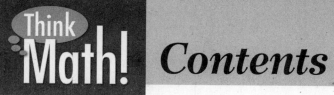

Contents

 **Contents**

Contents

These pages provide review of previously learned skills and concepts. The Spiral Review Book pages provide a comprehensive overview of math skills taught in the program.

Spiral Review Book

● Measurement

What time is it?

1.

```
:00
```

2.

```
:00
```

3.

```
:00
```

4.

```
:00
```

Number and Operations

What is the sum?

5. 3 + 3 = _____

6. 5 + 5 = _____

7. 10 + 10 = _____

8. 9 + 9 = _____

9. 7 + 7 = _____

10. 6 + 6 = _____

Geometry

Match each figure to its name.

1.

   sphere

2.

   rectangular prism

3.

   cylinder

Number and Operations

What number is shown?

4.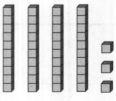

   _____

5.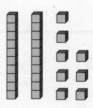

   _____

6.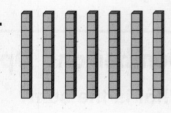

   _____

7.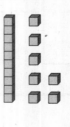

   _____

8.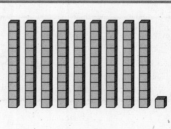

   _____

9.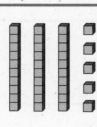

   _____

● Algebra

What comes next? Continue each pattern.

1.

_____ _____ _____

2.

_____ _____ _____

3.

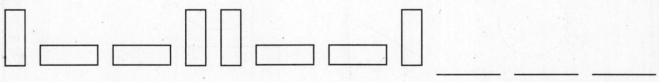

_____ _____ _____ _____

4.

1, 2, 3, 1, 2, 3, 1, 2, 3, _____, _____, _____, _____, _____

5.

1, 3, 5, 7, 9, 11, _____, _____, _____, _____, _____

6.

2, 4, 6, 8, 10, _____, _____, _____, _____, _____

Geometry

Follow the directions to color the figures. Then count each kind of figure and write the number.

1. Color the triangles green. _____

2. Color the circles red. _____

3. Color the squares blue. _____

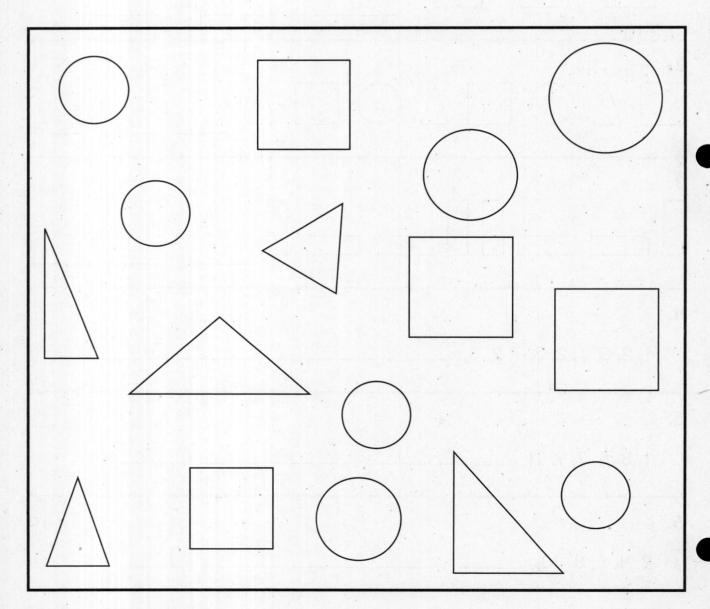

© Education Development Center, Inc.

● Measurement

What time is it?

1.  I hour later 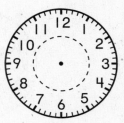 I hour later

| :00 | :00 | :00 |

Number and Operations

What is each sum?

● **2.** 5 + 5 = _____   5 + 6 = _____

3. 6 + 6 = _____   6 + 7 = _____

4. 7 + 7 = _____   7 + 8 = _____

5. 8 + 8 = _____   8 + 9 = _____

6. 9 + 9 = _____   9 + 10 = _____

● **7.** 10 + 10 = _____   10 + 11 = _____

Measurement

What time is it?

1.

2 hours
later

[|0:00]

[:00]

2.

2 hours
later

[:00]

[:00]

3.

2 hours
later

[:00]

[:00]

● Algebra

Skip-count. Continue each pattern.

1. 10, 9, 8, _____, _____, _____, _____, _____

2. 25, 24, 23, _____, _____, _____, _____, _____

3. 25, 23, 21, _____, _____, _____, _____, _____

4. 0, 5, 10, _____, _____, _____, _____, _____

5. 0, 10, 20, _____, _____, _____, _____, _____

6. 1, 11, 21, _____, _____, _____, _____, _____

Number and Operations

Write a number sentence about each picture.

7.

 6 + ☐ = ☐

8.

9 + ☐ = ☐

9.

 ☐ + 3 = ☐

10.

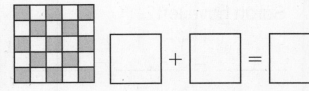 ☐ + ☐ = ☐

Number and Operations

How many coins are there? What is the value?

1.

_____ coins

_____ ¢

2.

_____ coins

_____ ¢

3.

_____ coins

_____ ¢

4.

_____ coins

_____ ¢

Problem Solving

Write a number sentence to describe each problem.

5. Sarah collected 8 shells at the beach. She gave 4 shells to her sister. How many shells does Sarah have left?

_____ − _____ = _____

6. Josh has 3 white shells and 6 striped shells. How many shells does he have altogether?

_____ + _____ = _____

● **Measurement**

What time is it?

1.

8:00

2.

:

3.

:

4.

:

5.

:

6.

:

7.

:

8.

:

Name _____ Date _____

Number and Operations

Which pairs make 10? Circle them as fast as you can.

2 8	3 6	3 7	8 2	9 1	6 3	2 9	5 6
4 6	6 4	6 5	8 3	9 1	1 8	1 9	2 7
8 2	9 2	4 6	5 6	4 7	3 6	7 3	2 8
4 6	5 5	6 3	7 3	2 8	2 9	7 2	7 3
8 2	3 8	4 7	5 5	1 8	2 9	3 7	4 6
5 5	6 4	7 3	8 3	9 2	1 9	3 7	5 5

Chapter 1

● Data Analysis and Probability

1. How many children chose pink as their favorite color?

_____ children

Number and Operations

How many fingers are up? How many are down?

2.

_____ up

_____ down

3.

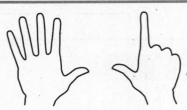

_____ up

_____ down

4.

_____ up

_____ down

5.

_____ up

_____ down

Measurement

Which weighs more?

1.

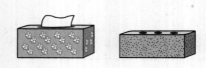

tissues   brick

2.

paper   baseball

3.

soup   pencil

Algebra

What is each fact family?

4.

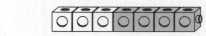

_3_ + _____ = 7   7 − _____ = _____

_____ + _____ = 7   7 − _____ = _____

5.

_____ + _____ = 10   10 − _____ = _____

_____ + _____ = 10   10 − _____ = _____

6.

_____ + _____ = 8   8 − _____ = _____

_____ + _____ = 8   8 − _____ = _____

● **Algebra**

Skip-count. What are the missing numbers?

1.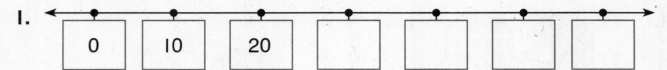

| 0 | 10 | 20 | | | | |

2.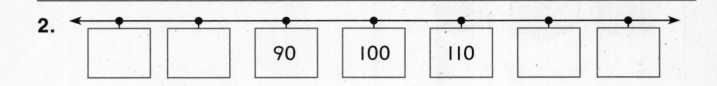

| | | 90 | 100 | 110 | | |

3.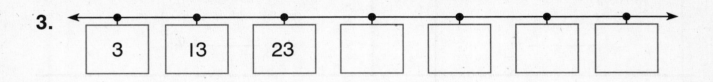

| 3 | 13 | 23 | | | | |

●

4.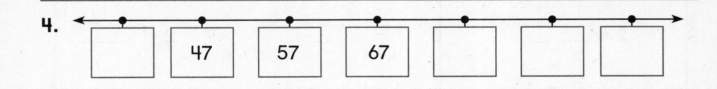

| | 47 | 57 | 67 | | | |

Geometry

What is the best name for each group?

5.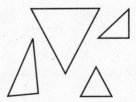

6.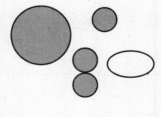

7.

_____   _____   _____

●

Number and Operations

How many coins are there? What is the value?

1.

_____ coins

_____ ¢

2.

_____ coins

_____ ¢

What is each sum?

3. $7 + 2 =$ _____ $8 + 2 =$ _____ $9 + 2 =$ _____

4. $6 + 3 =$ _____ $6 + 4 =$ _____ $6 + 5 =$ _____

5. $2 + 7 =$ _____ $3 + 7 =$ _____ $4 + 7 =$ _____

6. $8 + 1 =$ _____ $9 + 1 =$ _____ $10 + 1 =$ _____

Algebra

Skip-count. What comes next?

1.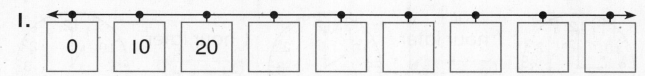

| 0 | 10 | 20 | | | | | | |

2.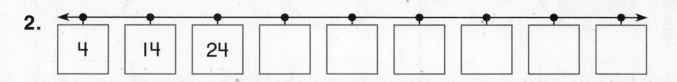

| 4 | 14 | 24 | | | | | |

Number and Operations

Circle number pairs with a sum of 10.

3.

| 4 | 7 | 2 | 5 | 9 |
| 6 | 3 | 8 | 6 | 2 |

4.

| 4 | 7 | 3 | 6 | 9 |
| 7 | 2 | 8 | 5 | 1 |

Problem Solving

5. The neighborhood has 5 streets that go North-South.
Each street crosses 3 avenues that go East-West.
How many intersections are there?

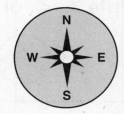

_____ intersections

Measurement

What time is it?

1.

 I hour later  I hour later

| :00 | | :00 | | :00 |

2.

 I hour later  I hour later

| :30 | | :30 | | :30 |

Algebra

Write >, <, or =.

3.   9 $\bigcirc$ 10 | **4.**   10 $\bigcirc$ 2 + 8 | **5.** 4 + 6 $\bigcirc$ 4 + 5

6. 7 + 3 $\bigcirc$ 10 | **7.** 5 + 5 $\bigcirc$ 6 + 5 | **8.** 9 + 1 $\bigcirc$ 1 + 8

● Number and Operations

Use patterns to add and subtract 10.

1. $3 + 10 = 13$

    $10 + 10 = \boxed{}$

    $21 + 10 = \boxed{}$

    $\boxed{} + 10 = 32$

    $\boxed{} + 10 = 42$

2. $\boxed{} - 10 = 30$

    $50 - 10 = \boxed{}$

    $53 - 10 = \boxed{}$

    $\boxed{} - 10 = 47$

    $\boxed{} - 10 = 57$

Find each missing number.

3. $3 + 7 = 10$

    $\boxed{} + 1 = 10$

    $4 + \boxed{} = 10$

    $\boxed{} + 8 = 10$

4. $5 + \boxed{} = 10$

    $10 + \boxed{} = 10$

    $8 + \boxed{} = 10$

    $7 + \boxed{} = 10$

© Education Development Center, Inc.

Algebra

Compare the value of the groups.
Write >, <, or =.

1.

 ◯

2.

 ◯

3.

 ◯

4.

 ◯

Data Analysis and Probability

5. How many more children were absent
   on Monday than on Wednesday?   _____ children

Children Absent from School					
Monday	𝙓	𝙓	𝙓		
Tuesday	𝙓	𝙓			
Wednesday	𝙓				
	1	2	3	4	5

Key: Each 𝙓 stands for 1 absent student.

● Algebra

Write >, <, or =.

1. 17 ◯ 71

2. 60 ◯ 6

3. 100 ◯ 10

4. 32 ◯ 23

5. 8 ◯ 28

6. 90¢ ◯ $1.00

7. 3 + 3 ◯ 6

8. 1 hour ◯ 10 minutes

Measurement

What time is it?

9.

10.

11.

:

© Education Development Center, Inc.

Algebra

Skip-count. What is missing?

1.

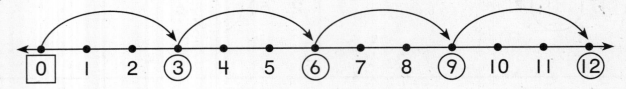

Number of Jumps	0	1	2	3	4				
Landing Number	0	3	6	9	12				

2.

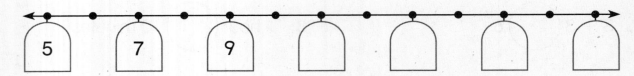

Number and Operations

Draw the jump. Complete the number sentence.

3.

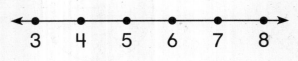

8 − ☐ = 5

4.

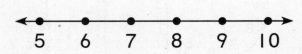

10 − ☐ = 6

© Education Development Center, Inc.

● Data Analysis and Probability

1. Draw the tallies.

Do you like to play soccer?				
yes	yes	yes	no	yes
no	yes	yes	yes	no

Do you like to play soccer?	
Yes	**No**

Number and Operations

What is the value?

2.

_____ ¢

3.

_____ ¢

Problem Solving

4. There are 10 birds in a tree. Then 3 of the birds fly away. How many birds are still in the tree?

_____ birds

5. There are 4 bugs on a leaf. Then 6 more bugs join them. How many bugs are on the leaf now?

_____ bugs

Algebra

Write >, <, or =.

1. $4 + 3$ ◯ $1 + 6$

2. $3 + 3$ ◯ $1 + 4$

Number and Operations

What numbers are missing?

3. $10 + 3 = \boxed{}$

  $20 + 3 = \boxed{}$

  $30 + 3 = \boxed{}$

  $40 + 3 = \boxed{}$

4. $10 + 8 = \boxed{}$

  $20 + 8 = \boxed{}$

  $30 + 8 = \boxed{}$

  $40 + 8 = \boxed{}$

5. $10 + 6 = \boxed{}$

  $20 + 6 = \boxed{}$

  $30 + 6 = \boxed{}$

  $40 + 6 = \boxed{}$

6. $10 + \boxed{} = 19$

  $\boxed{} + 9 = 29$

  $30 + \boxed{} = 39$

  $\boxed{} + 9 = 49$

7. $10 + \boxed{} = 15$

  $\boxed{} + 5 = 25$

  $\boxed{} + 5 = 35$

  $40 + \boxed{} = 45$

8. $10 + \boxed{} = 12$

  $\boxed{} + 2 = 22$

  $30 + \boxed{} = 32$

  $\boxed{} + 2 = 42$

● Geometry

What is the shape?

1. My shape is curved. It has no corners. It is flat.

_____

2. My shape has 4 sides. They are all the same length. It has 4 square corners.

_____

Number and Operations

What is the number?

3.   _____

4.   _____

5. 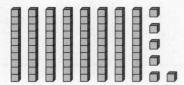 _____

6.   _____

7. 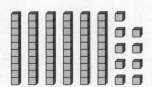 _____

8. 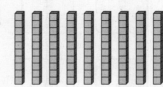 _____

Name _____ Date _____

Number and Operations

Which pairs make 10? Circle them
as fast as you can.

Sums of 10 Search

4 6	3 7	7 2	2 8	9 1	5 2	9 2	1 9	6 3
1 9	6 4	5 5	2 8	7 3	9 1	4 6	8 2	3 7
1 8	3 7	5 6	2 9	0 9	5 5	6 5	3 6	1 8
6 3	7 3	5 5	4 6	3 8	5 6	2 9	8 2	4 7
8 2	7 2	2 9	5 5	3 7	4 6	3 6	2 8	7 3
6 3	1 9	3 7	0 9	4 6	7 3	9 1	8 2	6 4

Chapter 3

● Measurement

What time is it? Follow the arrows.

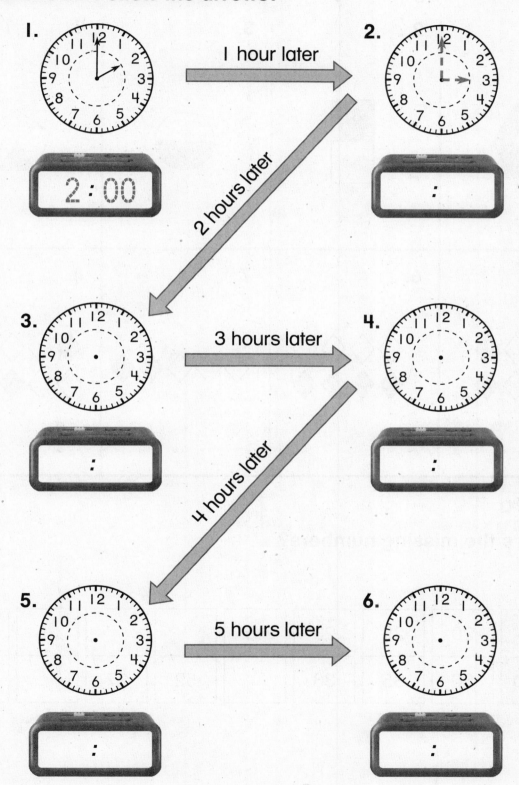

Number and Operations

Circle the drawings that show one half.

1.	2.	3.	4.

5.	6.	7.	8.
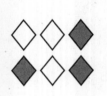		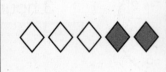	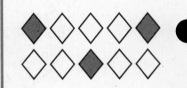

Algebra

What are the missing numbers?

9.

n	20	25	23	15			35	
$n + 10$	30	35	33		52	27		92

Name _____ Date _____

● Algebra

I. What comes next?

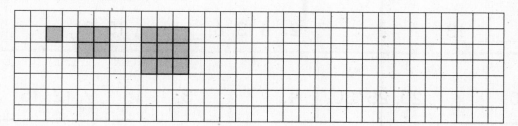

Measurement

What time is it?

2.

3.

Number and Operations

4. How can you show the same amount with the fewest coins? Draw (D) (P) (N) for coins.

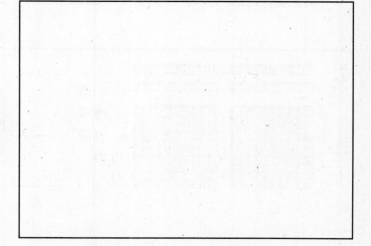

Algebra

What numbers are missing?

1. 1, 4, 7, _____, 13

2. 1, 3, 6, _____, 15, 21

3. _____, 10, _____, 20, 25, 30

4. _____, 3, _____, 9, 12, 15

Number and Operations

What is missing?

5. 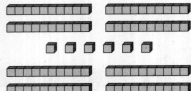 _____ hundreds _____ tens _____ ones

_____ + _____ + _____ = _____

6. 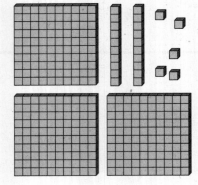 _____ hundreds _____ tens _____ ones

_____ + _____ + _____ = _____

7. 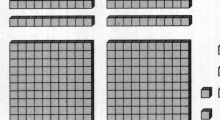 _____ hundreds _____ tens _____ ones

_____ + _____ + _____ = _____

● Number and Operations

How many coins are there? What is the value?

1.

_____ coins

_____ ¢

2.

_____ coins

_____ ¢

3. Circle two sets showing the same amount.

Problem Solving

4. Miguel ran for 7 minutes. He took a short break. Then he ran some more. If Miguel ran for 12 minutes all together, how long did he run after his break?

_____ minutes

Number and Operations

What is the order from smallest to biggest?

1.

| 93 | 29 | 75 | 48 | 121 |

_____ , _____ , _____ , _____ , _____

2.

| 136 | 163 | 13 | 36 | 61 |

_____ , _____ , _____ , _____ , _____

3.

| 18 | 180 | 81 | 118 | 108 |

_____ , _____ , _____ , _____ , _____

Measurement

What time is it?

4.

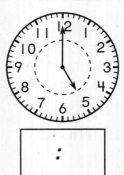

:

5.

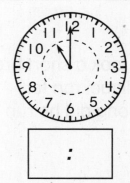

:

6.

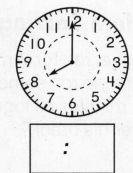

:

● Data Analysis and Probability

Is it certain, likely, or unlikely? Draw lines to match.

1. Your school will serve milk at lunch.   certain

2. All of your classmates will eat the
same dinner tonight.   likely

3. The sun will set tomorrow evening.   unlikely

Number and Operations

What is each difference?

4. $9 - 2 =$ _____	**5.** $11 - 3 =$ _____	**6.** $18 - 9 =$ _____
7. $10 - 4 =$ _____	**8.** $8 - 5 =$ _____	**9.** $12 - 4 =$ _____
10. $14 - 7 =$ _____	**11.** $10 - 5 =$ _____	**12.** $13 - 9 =$ _____
13. $7 - 6 =$ _____	**14.** $15 - 8 =$ _____	**15.** $11 - 6 =$ _____

Algebra

Look at the number on each line. Write the tens that sandwich the number. Which ten is closer?

1.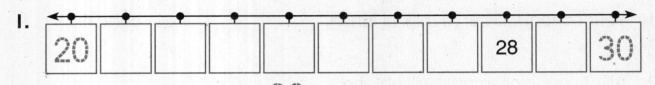

| 20 | | | | | | | | | 28 | | 30 |

The nearest ten to 28 is __30__.

2.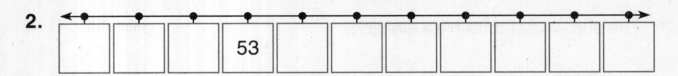

| | | | 53 | | | | | | | | |

The nearest ten to 53 is _____.

3.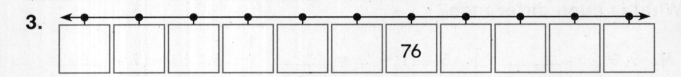

| | | | | | | 76 | | | | |

The nearest ten to 76 is _____.

Geometry

What figure is it? Match each word to a figure.

4. circle

5. triangle

6. square

● **Algebra**

Skip-count. What are the missing numbers?

1.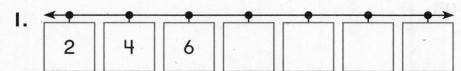

| 2 | 4 | 6 | | | | |

2.

| | | 15 | 20 | 25 | | |

3.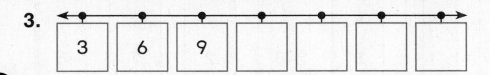

| 3 | 6 | 9 | | | | |

4.

| | 20 | 30 | 40 | | | |

Number and Operations

Write a number sentence about each picture.

5. ○○○○○
   ○○○○○
   ○○○○○

   _____ + _____ = _____

6.

   _____ + _____ = _____

Number and Operations

Which pairs make sums greater than 10? Circle them as fast as you can.

1.

| 7 3 | 7 4 | 5 7 | 4 4 | 4 6 | 5 6 | 8 2 | 3 9 | 1 9 |

2.

| 3 6 | 5 5 | 3 8 | 8 1 | 5 8 | 6 7 | 7 2 | 7 7 | 9 5 |

3.

| 4 8 | 2 7 | 9 4 | 9 0 | 1 9 | 2 9 | 1 8 | 4 4 | 6 6 |

Problem Solving

4. Sam has 15 pages left to read in his book. He reads 8 pages before dinner. How many pages does he have left to read?   _____ pages

5. Rachel had 20 minutes of reading homework and 15 minutes of math homework. How many minutes did Rachel spend on her homework altogether?   _____ minutes

⬤ Measurement

What time is it? Match the clocks.

1.

10:00

2.

2:00

3.

9:30

Number and Operations

What is missing? Complete each fact family.

4.

$7 + \_\_\_\_ = 12$

$5 + \_\_\_\_ = 12$

$12 - 5 = \_\_\_\_$

$12 - 7 = \_\_\_\_$

5.

$4 + \_\_\_\_ = 10$

$6 + \_\_\_\_ = 10$

$10 - 4 = \_\_\_\_$

$10 - 6 = \_\_\_\_$

Algebra

Write the numbers. Then write >, <, or =.

1. 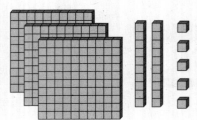 ◯

   _____   _____

---

2. 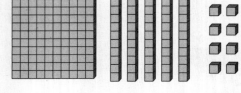 ◯

   _____   _____

Problem Solving

3. Tyra is making a necklace. She has beads in two shapes. If she continues the pattern, what will be the shape of the 18th bead? Explain.

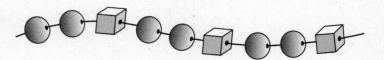

_____

_____

● Algebra

Use the pattern. Write the missing numbers.

1. $8 + 8 = 10 + 6$

$8 + 7 = 10 +$ _____

$8 + 6 = 10 +$ _____

$8 +$ _____ $= 10 + 3$

$8 +$ _____ $= 10 +$ _____

2. $9 + 9 = 10 + 8$

$9 + 8 = 10 +$ _____

$9 +$ _____ $= 10 + 6$

$9 +$ _____ $= 10 +$ _____

_____ $+$ _____ $=$ _____ $+$ _____

Number and Operations

● Use the number line to complete each addition sentence.

3.

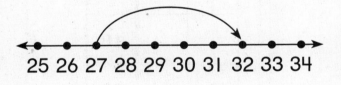

$27 +$ _____ $= 32$

4.

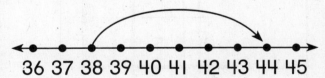

$38 + 6 =$ _____

5.

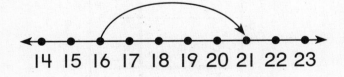

_____ $+ 5 = 21$

6.

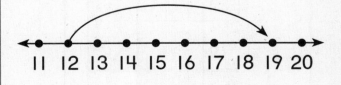

$12 +$ _____ $= 19$

© Education Development Center, Inc.

Number and Operations

Show the fewest coins for the money amount.
Draw (Q) for quarters, (D) for dimes, (N) for
nickels, and (P) for pennies.

1. 29¢

2. 43¢

What are the missing numbers? Amounts on
both sides of a thick line must be the same.

3.

9	3	
0	5	

4.

	1	
		18
3		20

5.

4		
8		10
	4	

6.

		8
3		
6	9	

© Education Development Center, Inc.

● Geometry

Draw the other half of each picture.

1.

2.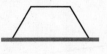

3.

Number and Operations

Complete each fact family.

4.

$3 + \underline{\hspace{1cm}} = 12$

_____

_____

5.

_____

_____

_____

_____

Measurement

About how long is each side in inches?

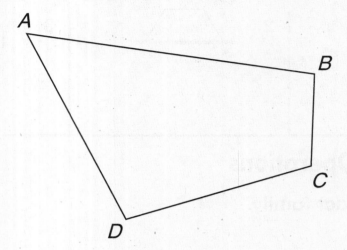

1. side *AB*	2. side *BC*	3. side *DC*	4. side *AD*
_____ inches	_____ inches	_____ inches	_____ inches

Data Analysis and Probability

A bag contains square and circle blocks. Which block do you think Sheli is more likely to pull out next from the bag?

5.

Square	Circle
ﾊﾊ ﾊﾊ ﾊﾊ lll	ll

_____

6.

Square	Circle
ﾊﾊ	ﾊﾊ ﾊﾊ ﾊﾊ

_____

● **Algebra**

Continue each pattern.

1.

_____ _____

2.

A B C D E F _____ _____ _____ _____ _____

3.

4.

1 2 2 3 3 3 4 4 4 4 _____ _____ _____ _____ _____

5.

1, 3, 6, 10, 15, 21, _____, _____, _____, _____, _____

6.

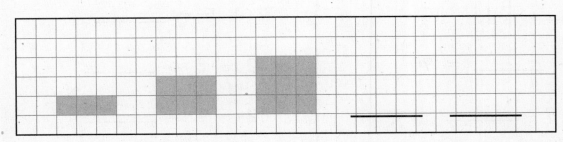

Geometry

Draw each figure.

1. I have 3 sides. All of
   my sides are different lengths.
   What figure am I?

   _____

2. I have 4 sides. All of
   my sides are the same length.
   What figure am I?

   _____

3. I have 4 sides and 4 square
   corners. I am NOT a square.
   What figure am I?

Number and Operations

What is missing? Complete each fact family.

4. $8 + 7 = 15$

   $7 + 8 = $ _____

   _____ $- 7 = 8$

   $15 - $ _____ $= 7$

5. $6 + 3 = 9$

   $3 + $ _____ $= 9$

   $9 - 6 = $ _____

   _____ $- 3 = 6$

6. _____ $+ 9 = 10$

   _____ $+ 1 = 10$

   $10 - $ _____ $= 9$

   $10 - 9 = $ _____

7. $5 + 7 = $ _____

   $7 + 5 = $ _____

   $12 - 5 = $ _____

   $12 - 7 = $ _____

● Number and Operations

What is the missing number?

1. 8 + 6 = _____

2. 10 + 20 = _____

3. 9 + 6 = _____

4. 20 + 20 = _____

5. 7 + 8 = _____

6. 40 + 30 = _____

7. 9 + 8 = _____

8. 50 + 10 = _____

● Data Analysis and Probability

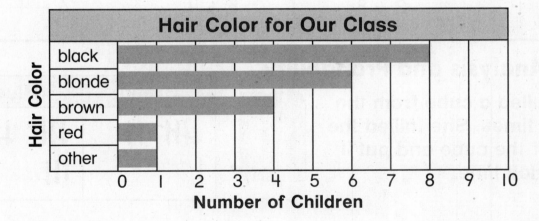

9. How many children have blonde hair? _____ children

10. How many more children have black hair than red hair? _____ children

Number and Operations

How can you make 15¢? Circle the coins.

1.

2.

3. Circle the facts with a sum of 10.

5 + 3	9 + 1	4 + 6	7 + 5
7 + 3	5 + 5	2 + 7	3 + 7
10 + 1	2 + 8	5 + 4	6 + 4

Data Analysis and Probability

Kate pulled a cube from the bag 20 times. She tallied the color of the cube and put it back each time.

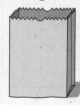

Green	Blue
卌 II	卌 卌 III

4. What colors are the cubes in the bag? _____

5. Which color is Kate less likely to pull out next? _____

6. Which color is Kate more likely to pull out next? _____

Algebra

What number is missing?

1. 4, 6, 8, _____, 12

2. 20, _____, 40, 50, 60

3. 21, 24, 27, 30, _____

4. _____, 50, 55, 60, 65

Measurement

About how many centimeters long is each object?

5.

about _____ centimeters

6.

about _____ centimeters

7.

about _____ centimeters

Number and Operations

What number matches each word?

1. seventeen .... 17 _____	2. eleven .... _____
3. twenty-four .... _____	4. thirty .... _____
5. ninety-three .... _____	6. eighty-one .... _____
7. sixty-two .... _____	8. one hundred .... _____
9. two hundred five ...... _____	10. two hundred fifty .... _____

Algebra

Write a digit to make each sentence true.

11. __3__ 6 > 26   12. _____ 4 < 58   13. _____ 2 > 86

14. 215 < _____ 15   15. 380 > _____ 80   16. 113 < _____ 23

17. 2 _____ 4 > 264   18. 3 _____ 1 < 350   19. 1 _____ 6 > 188

20. 438 < 4 _____ 1   21. _____ 77 > 194   22. 3 _____ 7 < 317

● Algebra

Look at the number on each line. Write the tens that sandwich the number. Which ten is closer?

1.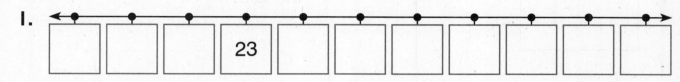

   The nearest ten to 23 is _____.

2.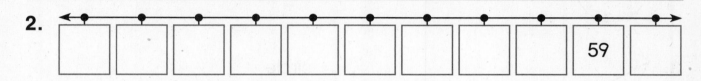

   The nearest ten to 59 is _____.

3.

   The nearest ten to 14 is _____.

4.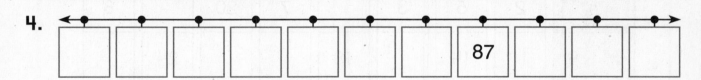

   The nearest ten to 87 is _____.

Problem Solving

5. Sarah has 5 marbles. A friend gives her 10 more. How many marbles does Sarah have now?

   _____ marbles

6. Kyle has 50 baseball cards. He gives 6 cards to a friend. How many cards does Kyle have now?

   _____ cards

Name _____ Date _____

Geometry

Draw lines to match each figure to its name.

1. 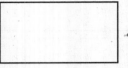   triangle

2.   rectangle

3. 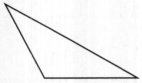   circle

Algebra

4. What are the missing numbers?

n	2	5	3		7	20		8
n + 9	11	14		18			24	

Number and Operations

5. What is the value of the 7 in 72? _____

6. What is the value of the 1 in 91? _____

7. What is the value of the 5 in 546? _____

● Number and Operations

Write each number.

1. 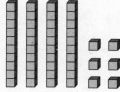 _____

2. _____

3.  _____

4. 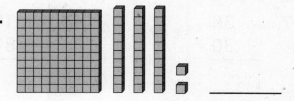 _____

5. 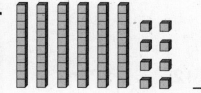 _____

6. 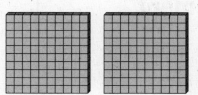 _____

Show each amount using the fewest coins.

Use (D) for dimes and (P) for pennies.

7. | 52¢ |

8. | 24¢ |

Number and Operations

Add.

1. 7 + 10 = _____	**2.** 20 + 10 = _____	**3.** 41 + 20 = _____
4. 50 + 50 = _____	**5.** 50 + 49 = _____	**6.** 22 + 80 = _____
7. $\begin{array}{r} 38 \\ + 30 \\ \hline \end{array}$	**8.** $\begin{array}{r} 10 \\ + 80 \\ \hline \end{array}$	**9.** $\begin{array}{r} 65 \\ + 60 \\ \hline \end{array}$

Measurement

What time is it?

10.

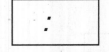

11.

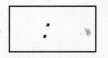

12.

13.

14.

15.

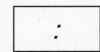

● Algebra

Write <, >, or =.

1. 8 ◯ 80

2. 10 ◯ 0

3. 19 ◯ 21

4. 67 ◯ 76

5. 4 + 5 ◯ 8

6. 20 ◯ 10 + 10

7. 2 + 2 ◯ 3 + 4

8. 7 + 10 ◯ 8 + 8

Number and Operations

Subtract.

9. 10 − 5 = _____

10. 9 − 9 = _____

11. 15 − 5 = _____

12. 20 − 10 = _____

13. 12 − 6 = _____

14. 30 − 5 = _____

Algebra

What is missing? Continue each pattern.

I. 6, 8, 10, 12, 14, 16, _____, _____, _____, _____, _____

_____

2. 5, 9, 13, 17, 21, 25, _____, _____, _____, _____, _____

_____

3. 100, 90, 80, 70, 60, 50, _____, _____, _____, _____, _____

Number and Operations

What is missing? Complete each puzzle.

4.

5	7	12
		12
11	13	

5.

	5	15
10	15	
	20	40

6.

11	8	
	9	12
14		31

7.

20		40
50		60
70	30	

● Algebra

What are the missing numbers?

1. $4 +$ _____ $= 8$

2. $0 +$ _____ $= 9$

3. _____ $+ 2 = 7$

4. $8 +$ _____ $= 13$

5. $15 +$ _____ $= 16$

6. _____ $+ 4 = 20$

7. _____ ¢ $+ 20$¢ $= 70$¢

8. _____ ¢ $+ 2$¢ $= 82$¢

9. _____ $- 8 = 1$

10. $14 -$ _____ $= 3$

11. $20 -$ _____ $= 16$

12. _____ $- 5 = 5$

13. $18 -$ _____ $= 9$

14. _____ $- 10 = 20$

15. $\$1.00 -$ _____ ¢ $= 50$¢

16. _____ ¢ $- 10$¢ $= 60$¢

Problem Solving

A plane travels 10 miles in 1 minute. It takes off at 5:00.

17. How far has the plane traveled by 5:10?

_____ miles

18. What time will it be when the plane has traveled 50 miles?

_____ : _____

Measurement

What is the length to the nearest inch?

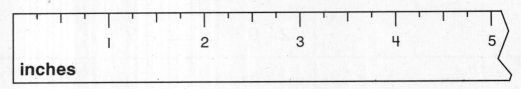

1.

about _____ inches

2.

about _____ inches

---

3. What is the length of the crayon to the nearest centimeter?

about _____ centimeters

● Geometry

Draw a line of symmetry for each figure.

1.

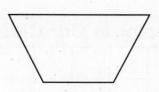

2.

3.

4.

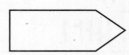

● Number and Operations

Complete each fact family.

5. $6 + 5 = 11$

$11 - 5 = 6$

_____ + _____ = _____

_____ − _____ = _____

6. $8 + 12 = 20$

$20 - 12 = 8$

_____ + _____ = _____

_____ − _____ = _____

7. $7 + 9 = 16$

$16 - 7 = 9$

_____ + _____ = _____

_____ − _____ = _____

8. $19 + 0 = 19$

$19 - 0 = 19$

_____ + _____ = _____

_____ − _____ = _____

Data Analysis and Probability

1. Use the table to make a pictograph.
   Choose a symbol and make a key.

Hair Color in Our Class	
Color	**Tally**
Blonde	⊥⊥⊥⊥ |||
Red	||
Brown	⊥⊥⊥⊥	
Black	⊥⊥⊥⊥ |||

Hair Color in Our Class							
Blonde							
Red							
Brown							
Black							

Key: Each _____ stands for _____ children.

2. How many more children have black
   hair than red hair?   _____ children

3. How many children are in the class?   _____ children

Number and Operations

What is the value?

4.

_____ ¢

5.

_____ ¢

● Geometry

Color the figures that are congruent to the first one.

1.

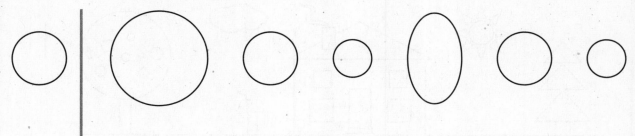

2.

Problem Solving

Solve.

3. Pete leaves home at 8:10. It takes him 20 minutes to walk to school. When does he get to school?

   _____:_____

4. Kim wants to buy an apple for 35¢. She has 2 dimes and 1 nickel. How much more money does she need?

   _____ ¢

5. Betsy took 6 books out of the library. She read half of them. How many books did she read?

   _____ books

6. Walt has a piece of wood. It is 4 feet long. He cuts it in 2 equal pieces. How long is each piece of wood?

   _____ feet

Geometry

1. How many of each figure are there?

_____ circles   _____ triangles   _____ rectangles

Number and Operations

2. Which two sets show the same number? Circle them.

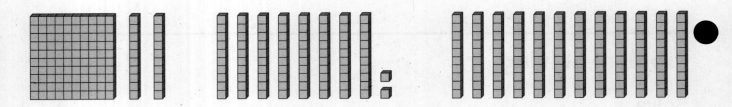

Algebra

Compare the sums. Write >, <, or =.

3. $6 + 3$ ◯ $5 + 4$

4. $7 + 2$ ◯ $5 + 3$

5. $2 + 5$ ◯ $4 + 4$

6. $8 + 1$ ◯ $4 + 6$

7. $1 + 6$ ◯ $2 + 4$

8. $3 + 5$ ◯ $6 + 2$

● Data Analysis and Probability

1. Which fruit did most
   of the children choose?

   _____

2. Which fruit did the fewest
   children choose?

   _____

3. How many children chose oranges?

   _____ children

Favorite Fruit of Second Graders	
apples	✝✝✝✝ |||
oranges	✝✝✝✝ ||
bananas	✝✝✝✝ ✝✝✝✝ ||||
grapes	✝✝✝✝
cherries	||

4. How many more children chose apples than grapes? _____ children

● 5. How many children answered the survey? _____ children

Number and Operations

Subtract.

6.
$$24 - 4$$
$$24 - 5$$

7.
$$48 - 8$$
$$48 - 9$$

8.
$$32 - 2$$
$$32 - 3$$

9.
$$65 - 5$$
$$65 - 6$$

Name _____ Date _____

Lesson 3

Measurement

What time is it?

1.	2.	3.	4.

Number and Operations

What is missing?

	Whole	Half
5.	12	6
6.	24	
7.		24
8.	50	
9.	52	

© Education Development Center, Inc.

SR60 Spiral Review Book

Chapter 8

● Algebra

What is missing? Continue each pattern.

1. 7, 10, 13, 16, 19, _____, _____, _____, _____, _____

2. 1, 10, 19, 28, 37, 46, _____, _____, _____, _____, _____

3. 90, 85, 80, 75, 70, _____, _____, _____, _____, _____

Number and Operations

Put the numbers in order from smallest to biggest.

● **4.**

| 25 | 205 | 112 | 86 | 147 |

_____, _____, _____, _____, _____

5.

| 130 | 59 | 47 | 128 | 75 |

_____, _____, _____, _____, _____

6.

| 310 | 290 | 106 | 95 | 121 |

●

_____, _____, _____, _____, _____

Algebra

Look at the number on each line. Write the tens that sandwich the number. Which ten is closer?

1.

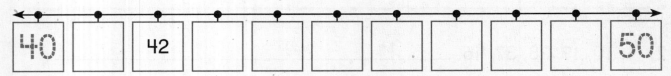

| 40 | | 42 | | | | | | | | 50 |

The nearest ten to 42 is _____.

2.

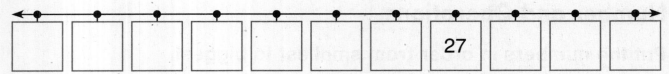

| | | | | | | | 27 | | | |

The nearest ten to 27 is _____.

3.

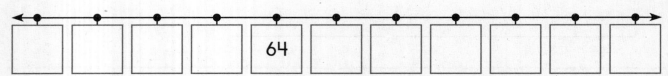

| | | | | 64 | | | | | | |

The nearest ten to 64 is _____.

Measurement

About what time is it? Circle the closer time.

4.

7:00   7:30

5.

11:30   12:00

6.

2:30   3:00

● Number and Operations

Subtract 8 from each number.

1.

28	29	88	89	48	49	39	18	78
20								

2.

10	20	30	50	90	91	92	93	94

3.

28	27	78	77	58	57	37	17	97

How many coins are there? What is the value?

4.

_____ coins

_____ ¢

5.

_____ coins

_____ ¢

Measurement

What time is it?

1.

2.

3.

   I hour later →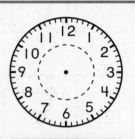

Number and Operations

Circle each picture with one half colored gray.

4.

5.

6.

7.

8.

9.

● Number and Operations

What is each sum?

| 1. $\begin{array}{r} 15 \\ + 5 \\ \hline \end{array}$ | 2. $\begin{array}{r} 5 \\ + 20 \\ \hline \end{array}$ | 3. $\begin{array}{r} 25 \\ + 5 \\ \hline \end{array}$ | 4. $\begin{array}{r} 5 \\ + 30 \\ \hline \end{array}$ |

5. $5 + 40 = \boxed{}$

6. $50 + 10 = \boxed{}$

7. $5 + 45 = \boxed{}$

8. $80 + 10 = \boxed{}$

● Problem Solving

What number sentence describes each problem?

9. Chris had 24 rocks in his collection. He found 8 more rocks at the park. How many rocks did he have in all?

_____

10. Clara has 15 brown rocks and 9 black rocks. How many more brown rocks than black rocks does Clara have?

_____

11. There were 35 people waiting in line for the movie. 9 more people joined the line. Then how many people were in the line?

_____

12. There were 27 people on a bus. 8 people got off at the next stop. How many people were on the bus then?

_____

Data Analysis and Probability

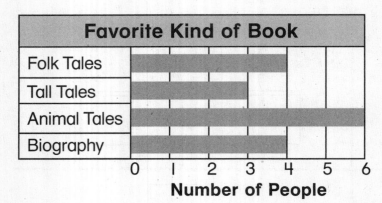

Favorite Kind of Book

Folk Tales	
Tall Tales	
Animal Tales	
Biography	

0 1 2 3 4 5 6
Number of People

1. How many people voted for Tall Tales?   _____ people

_____

2. Which kind of story got the most votes?   _____

_____

3. Which kind of story got the fewest votes?   _____

_____

4. How many more votes did Animal Tales get
   than Folk Tales?   _____ more votes

_____

5. How many people voted altogether?   _____ people

Number and Operations

What is missing? Complete each puzzle.

6.

70		84
	7	37

7.

	8	68
50		55

● Number and Operations

How much is half?

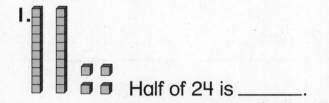

1. Half of 24 is _____.

2. Half of 16 is _____.

3. 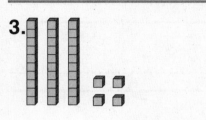 Half of 34 is _____.

4. 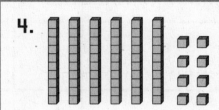 Half of 68 is _____.

Measurement

● What is missing?

5.  5 minutes later

6.  15 minutes later

© Education Development Center, Inc.

Number and Operations

What number matches each number word?

1. fifteen 15	2. nine .... ____
3. twenty .... ____	4. thirty-one .... ____
5. fifty-six .... ____	6. forty-two .... ____
7. seventy-three .... ____	8. one hundred seven .... ____
9. one hundred sixty ..... ____	10. two hundred one .... ____

How can you show each money amount with the fewest coins?

Money Amount	How many dimes?	How many pennies?	How much money?
11. twelve cents	1	2	12¢
12. eighteen cents			¢
13. twenty-three cents			¢
14. thirty cents			¢
15. forty-five cents			¢

● Measurement

Which object is longer in each pair? Circle it.
Mark an X on the shorter object.

1.

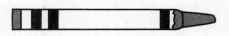

2.

3.

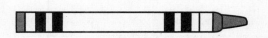

4.

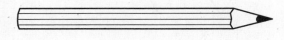

●

Number and Operations

What numbers are missing?

5.

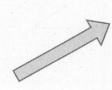

```
  12
+  8
```
□

```
+  8
———
  28
```

```
  28
+  8
```
□

6.

●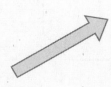

```
  64
-  8
```
□

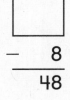

```
-  8
———
  48
```

```
  48
-  8
```
□

© Education Development Center, Inc.

Algebra

What is the value of each shape?

1.

☆ + ☆ = 20

☆ = _____

2.

⬡ + ⬡ = 18

⬡ = _____

3.

△ + △ = 32

△ = _____

4.

◇ + ◇ = 50

◇ = _____

Geometry

How many sides and corners are there?

5.

_____ sides

_____ corners

6.

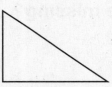

_____ sides

_____ corners

7.

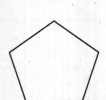

_____ sides

_____ corners

8.

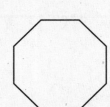

_____ sides

_____ corners

● Number and Operations

How can you show each money amount with the fewest coins?

	Money Amount	Number of Dimes	Number of Pennies	Total
1.	thirty-two cents	3	2	32¢
2.	eighty-six cents			
3.	forty cents			
4.	fifteen cents			
5.	nine cents			

What is each sum?

6. 60
   +10

7. 30
   +64

8. 80
   +20

9. 47
   +20

10. 35
   +30

11. 20
   +32

Number and Operations

What is each sum?

1.	2.	3.	4.
5 +15	7 +12	11 + 2	18 + 2

5.	6.	7.	8.
20 +30	39 +40	10 +57	24 +60

9. 24 + 24 = _____   **10.** 40 + 20 = _____

11. 16 + 14 = _____   **12.** 39 + 39 = _____

Problem Solving

Solve each problem.

13. Lisa bought a pencil for 15¢ and a marker for 45¢. How much did Lisa spend in all?

_____¢

14. Jay bought a ruler for 28¢. He gave the clerk 50¢. How much change did Jay get?

_____¢

● **Algebra**

Write >, <, or =.

1.	3 + 3 ◯ 4 + 3	**2.**	8 + 1 ◯ 5 + 4
3.	5 + 5 ◯ 6 + 2	**4.**	13 + 9 + 1 ◯ 10 + 11 + 2
5.	15 − 7 ◯ 17 − 7	**6.**	22 − 19 ◯ 22 − 9
7.	8 − 1 ◯ 9 − 2	**8.**	40 − 14 ◯ 27 − 14

●

What numbers are missing from each table?
Write the rule.

9.

2	4
3	6
4	
5	10
6	
	14

10.

4	9
5	11
10	21
3	
11	
9	

●

Rule: _____   Rule: _____

Data Analysis and Probability

1. Use the tally table to complete the bar graph.

Favorite Ice Cream Flavor for Mr. Kim's Class	
vanilla	卌 II
chocolate	卌 卌 I
strawberry	IIII
mint	卌 III

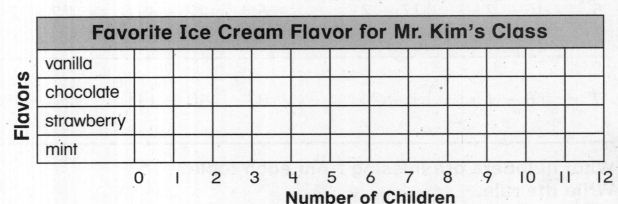

Favorite Ice Cream Flavor for Mr. Kim's Class												
vanilla												
chocolate												
strawberry												
mint												

Flavors

0 1 2 3 4 5 6 7 8 9 10 11 12
Number of Children

2. Which flavor is the favorite of the most children? _____

3. How many more children chose mint ice cream than strawberry ice cream? _____ children

4. Which flavor was chosen by the fewest children? _____

5. If every child in Mr. Kim's class chose a favorite flavor, how many children are in the class? _____ children

● Algebra

What comes next? Continue each pattern.

1.

_____ _____ _____

2. 6, 12, 18, 24, _____, _____, _____, _____, _____

3. 8, 11, 14, 17, 20, _____, _____, _____, _____, _____

● Number and Operations

Show the fewest coins for each amount of money. Draw Ⓠ for quarters, Ⓓ for dimes, Ⓝ for nickels, and Ⓟ for pennies.

4. 34¢

5. 28¢

6. 82¢

7. 56¢

Measurement

I. What time is it? Draw the clock hands.

_____ 6:00 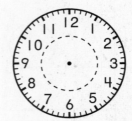 4 hours later _____ :

Number and Operations

How much money? Write >, <, or =.

2.

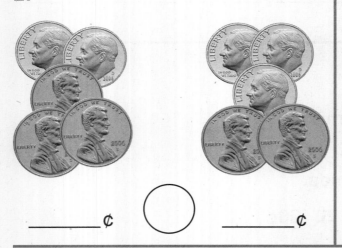

_____ ¢ ◯ _____ ¢

3.

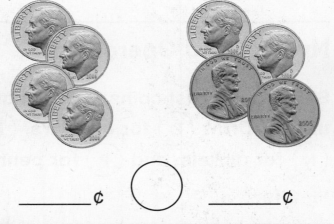

_____ ¢ ◯ _____ ¢

4. Which pairs make 10? Circle them as fast as you can.

6	4	2	0	3	2	5	8	1
3	6	8	9	7	9	5	3	9

4	1	7	8	6	9	7	4	5
7	8	2	2	4	1	3	5	6

● Number and Operations

What numbers are missing?

1.

9	5	
	3	9
15	8	

2.

8		18
4	6	
	16	28

3.

15	5	
	4	10
21		

4.

	4	
2		5
14	7	

5.

11	8	
9		9

6.

		18
	6	12
9		

7.

21		23
	2	27

8.

		19
	5	7
14		

9.

7		
	9	17
		25

Data Analysis and Probability

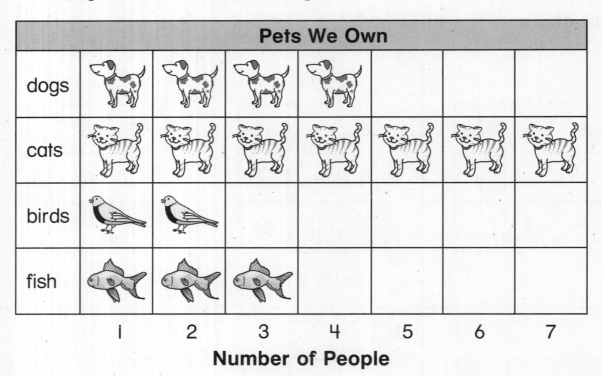

Pets We Own							
dogs	🐕	🐕	🐕	🐕			
cats	🐈	🐈	🐈	🐈	🐈	🐈	🐈
birds	🐦	🐦					
fish	🐟	🐟	🐟				
	1	2	3	4	5	6	7

Number of People

1. How many people have a dog? _____ people

2. Which pet is owned by more people than any other pet? _____

3. Which pet is owned by the fewest people? _____

4. How many more people own dogs than birds? _____ more people

Geometry

Circle the figure that does NOT belong in each group?

5.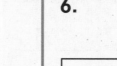

6.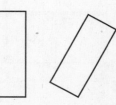

● Measurement

What time is it?

1.

2.

3.

4.

5.

6.

Number and Operations

**What is missing? Complete each puzzle so the
values on each side of the thick line are the same.**

7.

8.

Number and Operations

Show each amount of money with the fewest

coins. Draw , , , and 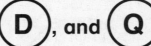 for the coins.

1. 50¢

2. 46¢

3. 87¢

4. 39¢

Algebra

Make each sentence true. Write >, <, or =.

5. 53 + 22 ◯ 54 + 22

6. 75 + 7 ◯ 75 + 4

7. 21 − 5 ◯ 21 − 10

8. 38 − 36 ◯ 48 − 46

● Number and Operations

Break the numbers in different ways so they are easier to add. Write the missing numbers.

1.

| 16 $+\ 8$ | ⇒ | $10 + 6$ $+\ 0 + 8$ ___ $+$ | ⇒ | $5 + 11$ $+\ 5 +\ \ 3$ ___ $+$ | ⇒ | $20 - 4$ $+\ 10 - 2$ ___ $-$ |

2.

| 25 $+19$ | ⇒ | $+\ 15$ $+14 +$ ___ $+$ | ⇒ | $20 +$ $+10 +$ ___ $+$ | ⇒ | $30 -$ $+\ \ \ - 1$ ___ $-$ |

3.

| 15 $+17$ | ⇒ | $+\ 5$ $+10 +$ ___ $+$ | ⇒ | $5 +$ $+\ 5 +$ ___ $+$ | ⇒ | $20 -$ $+\ \ \ - 3$ ___ $-$ |

Problem Solving

Write a number sentence to describe each problem.

4. Adam has 21 computer games. He buys 3 more. Now how many games does he have?

_____

5. Matt has 15 video games and 24 computer games. How many more computer games than video games does he have?

_____

Measurement

What is the area? Each ⬜ is 1 square unit.

1.

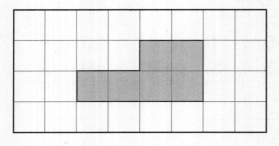

_____ square units

2.

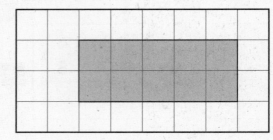

_____ square units

3.

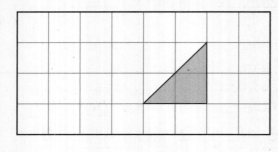

_____ square units

4.

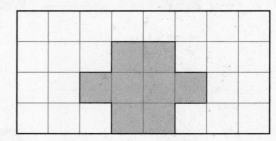

_____ square units

Number and Operations

What is missing? Draw the symbols and write the numbers.

5.

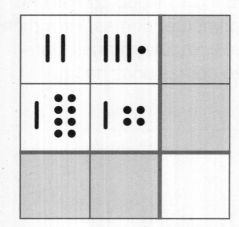

● Algebra

Make each sentence true. Write >, <, or =.

1. 45 + 22 ◯ 22 + 45

2. 76 + 18 ◯ 76 + 35

3. 55 − 8 ◯ 55 − 48

4. 84 − 29 ◯ 85 − 29

5. 45 + 43 ◯ 44 + 39

6. 25 + 25 ◯ 100 − 50

Measurement

What landmark on the clock can help you find the time? Write the landmark and the exact time.

7.

landmark time ____:____

exact time ____:____

8.

landmark time ____:____

exact time ____:____

Geometry

Draw the reflection of each figure.

1.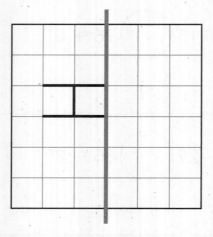

2.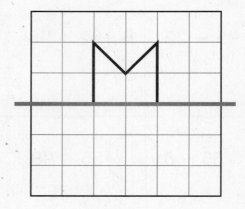

Number and Operations

What is missing?

3.

40		45
	9	19

→

	4	5
+	1	9

4.

20		26
	2	62

→

	2	6
+	6	2

● Number and Operations

What part is shaded? Write the fraction.

1.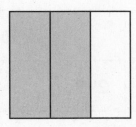

$\dfrac{2}{3}$
_____

2.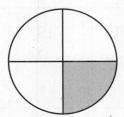

_____

3.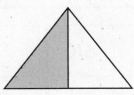

_____

4.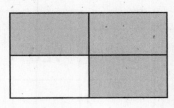

_____

● Algebra

Write the missing numbers. △ **can contain only ones from 0 to 9.** ☐ **can contain only tens from 0 to 90.**

5. $35 + 24 =$ [50] $+ 9 =$ ___59___

6. $43 + 21 = 60 +$ △ $=$ _____

7. $86 + 12 =$ ☐ $+$ △ $=$ _____

● 8. $64 + 23 =$ ☐ $+$ △ $=$ _____

Algebra

What is missing?

1.

n	450	719	391	72			246
n + 200					354	249	

2.

m	355	920	417	763		610	
m − 300					127		404

Number and Operations

3. Circle the number pairs with a sum of 100.
Work as fast as you can.

40 50	60 20	20 80	20 70	40 60	90 10	20 90	50 50	10 90
40 60	60 40	60 50	30 60	30 70	80 20	10 90	20 60	20 80

Chapter 10

● Data Analysis and Probability

Some second graders were asked to choose one favorite sport. Here are the results.

Favorite Sports of Second Graders	
baseball	卌 卌 ll
basketball	卌
football	卌 ll
soccer	卌 llll

1. Which sport was chosen most? _____

2. Which sport did the fewest children choose? _____

● 3. How many children chose soccer? _____

4. How many more children chose baseball than soccer? _____

5. How many children answered the survey? _____

Number and Operations

What is the sum?

6. $7 + 3 =$ _____

7. $27 + 3 =$ _____

8. $27 + 13 =$ _____

9. $37 + 13 =$ _____

10. $2 + 8 =$ _____

11. $22 + 8 =$ _____

● 12. $22 + 18 =$ _____

13. $32 + 18 =$ _____

Algebra

What comes next?

1.

 ___ ___

2.

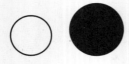 ___ ___ ___ ___

Number and Operations

What is missing?

3.

400	80	7	487
	10		512
	90	9	

4.

		2	202
	20		329
500		11	

5.

600		7	687
	30		134
		3	

6.

	90	8	398
200	70		275

● Number and Operations

Show each money amount as many ways as you can. Use only dimes and pennies.

1.

28¢	
Dimes	**Pennies**
0	28

How many coins are used to show 28¢ with the fewest dimes and pennies?

_____ coins

2.

31¢	
Dimes	**Pennies**

How many coins are used to show 31¢ with the fewest dimes and pennies?

_____ coins

Problem Solving

3. Claire has soccer practice at 4:00. Practice is one and a half hours long. What time will soccer practice be over?

_____

4. A movie starts at 7:30. It is over at 9:30. How long is the movie?

_____

5. Jim has $1.45. What are the fewest bills and coins he could have?

_____

6. Morgan buys a salad for $3.75. She pays $5.00. How much is her change?

$ _____

Measurement

1. What is missing?

5 minutes later
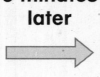

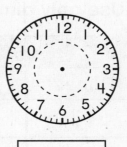

:

:

Number and Operations

What is missing?

2.

30		45
	5	20

3.

	50	12
36		6

4.

44		64
	6	22

5.

		9
62		
	60	14

● Data Analysis and Probability

1. At what time was it 48°?

_____ A.M.

2. How many degrees warmer was it at 11:00 A.M. than at 10:00 A.M.?

_____ degrees

3. How much did the temperature go up from 8:00 A.M. and 11:00 A.M.?

_____ degrees

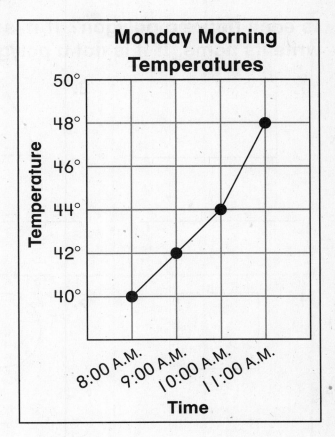

Problem Solving

Draw a picture to solve the problem.

4. Four children stand in a line from shortest to tallest. Dena is taller than Clara. Tim is the shortest. Dena is NOT the tallest. Where is Evan in the line?

_____

Geometry

**Is each figure a polygon? If it is a polygon,
write its name. If it is not a polygon, write *no*.**

1.

_____

2.

_____

3.

_____

4.

_____

5.

_____

6.

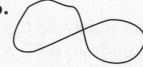

_____

Number and Operations

Show each amount with the fewest coins.

	Money Amount	Dimes	Nickels	Pennies	Total
7.	seventy-five cents	7	1	0	75¢
8.	eighty-eight cents				88¢
9.	ninety cents				90¢
10.	ninety-six cents				96¢

Number and Operations

Add or subtract.

1. $5 + 5 =$ _____

2. $10 - 5 =$ _____

3. $10 + 10 =$ _____

4. $30 - 15 =$ _____

5. $13 + 13 =$ _____

6. $42 - 21 =$ _____

7. $24 + 24 =$ _____

8. $50 - 25 =$ _____

9. $36 + 36 =$ _____

10. $92 - 46 =$ _____

Problem Solving

11. Sally has 2 red stickers, 6 yellow stickers, and 1 green sticker. How many stickers does she have in all?

_____ stickers

12. Jordan has a baseball card collection. He buys 30 new baseball cards. Now Jordan has 64 baseball cards. How many baseball cards did Jordan have before he bought new cards?

_____ cards

13. Nancy has 12 marbles. Half of them are striped. The rest are solid. How many marbles are solid?

_____ solid marbles

Measurement

March 2007						
Sunday	Monday	Tuesday	Wednesday	Thursday	Friday	Saturday
				1	2	3
4	5	6	7	8	9	10
11	12	13	14	15	16	17
18	19	20	21	22	23	24
25	26	27	28	29	30	31

1. How many days are in March? _____
2. Color the third Monday in the month red.
3. Color the first Thursday in the month blue.
4. Color the last Wednesday in the month yellow.

5. What day of the week is March 18th? _____

6. What day of the week is March 6? _____

Number and Operations

Add or subtract.

7. $\begin{array}{r} 425 \\ +507 \\ \hline \end{array}$ 8. $\begin{array}{r} 142 \\ +\ 36 \\ \hline \end{array}$ 9. $\begin{array}{r} 408 \\ -337 \\ \hline \end{array}$ 10. $\begin{array}{r} 318 \\ -152 \\ \hline \end{array}$

● Data Analysis and Probability

1. The table shows the favorite vegetables of Mrs. Taggert's class. Make a bar graph of the data.

Favorite Vegetables	
Type of Vegetable	**Number of Children**
Corn	4
Peas	3
Broccoli	5
Carrots	6
Spinach	2

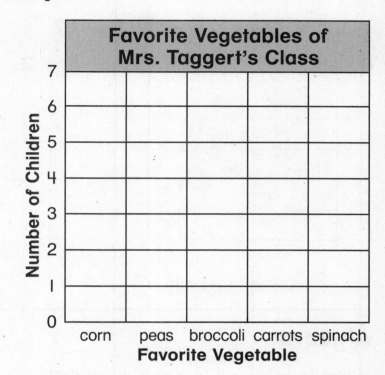

2. How many more children like carrots than spinach?
_____ more children

3. How many children are in Mrs. Taggert's class?
_____ children

4. What vegetable is the favorite of the most children?
_____

Geometry

Are the figures congruent? Write *yes* or *no*.

5.
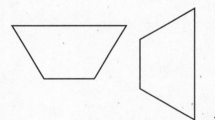
_____

6.
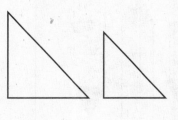
_____

Measurement

What is the area of each figure?

1.

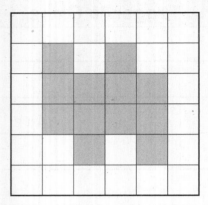

_____ square units

2.

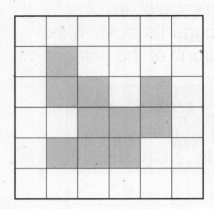

_____ square units

Algebra

What is missing in each pattern?

3.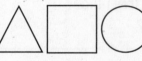

4.

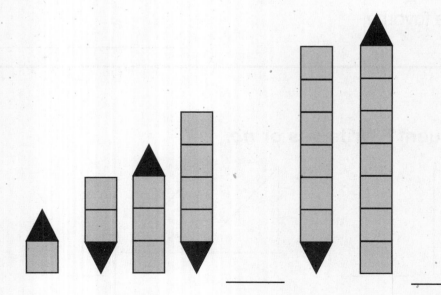

_____   _____

● **Number and Operations**

Complete the table.

┌─────────────────────────────────┐
│   **School Store**   │
│   │
│ eraser 7¢   crayon 9¢   │
└─────────────────────────────────┘

	Julia Has	She Buys	Total Cost	Change
1.	4 dimes	2 erasers, 2 crayons	_____¢	_____¢
2.	4 dimes	_____ erasers, 2 crayons	_____¢	1¢
3.	4 dimes	4 erasers, _____ crayons	28¢	_____¢

Draw what is missing.

4.

5.

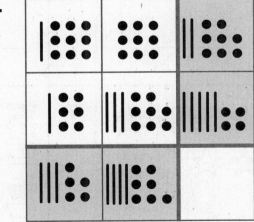

Algebra

How many are there?

1. How many books are in the bookcase?

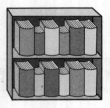

$2 \times 6 =$ _____ $6 \times 2 =$ _____

_____ books in all

2. How many muffins are in the pan?

$3 \times 4 =$ _____ $4 \times 3 =$ _____

_____ muffin in all

Number and Operations

What fraction is gray?

3.

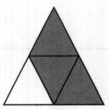

_____

4.

_____

5.

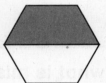

_____

6.

_____

7.

_____

8.

_____

● Number and Operations

What is the fraction for the shaded part?

1.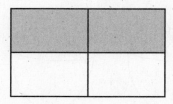

_____

2.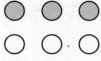

_____

3.

_____

4.

_____

Measurement

What time is it?

5.

_____

6.

_____

Algebra

What are the missing numbers for the doubles?

I.

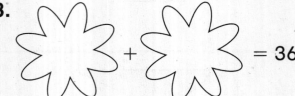

+ = 24

2.

+ = 68

3.

+ = 36

4.

+ = 120

Geometry

Draw a new figure. Double the sides of the gray figure.

5.

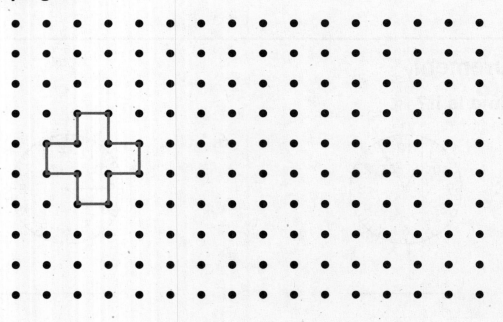

● Measurement

What is missing? Look for a pattern in the clocks.

1.

: _____

2.

: _____

3.

: _____

4.

: _____

5.

: _____

6.

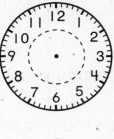

: _____

Algebra

7. What is missing?

n	1	2	3	5		15	18
$n + 20$	21	22		25	30	35	
$n + 22$	23	24	25		32		40

Data Analysis and Probability

The children took a survey. Each person chose one favorite sandwich.

Favorite Sandwich						
grilled cheese	🥪	🥪	🥪	🥪		
tuna	🥪					
peanut butter	🥪	🥪	🥪	🥪	🥪	🥪
turkey	🥪	🥪				

Key: Each 🥪 stands for 2 children.

I. How many children did NOT choose peanut butter? _____ children

2. How many more children chose peanut butter than turkey? _____ children

3. How many children took part in the survey? _____ children

Number and Operations

What is missing? Write the numbers.

4.

30		32
30	9	
60		

5.

50	8	
30		30
		88

Number and Operations

What is each amount in dollars and cents?

1.

_____

2.

_____

Problem Solving

3. Molly bought a domino set for $2.75. She paid for the set with $5.00. How much change did she get back?

_____

4. Dana has $6.50 in her bank. She wants to buy a book that costs $7.53. How much more money does Dana need?

_____

5. Lee bought a train car. He gave the clerk $5.00 and got back $1.15 in change. How much did the train car cost?

_____

6. Maya bought a toy dinosaur on sale for $2.95. The regular price was $4.65. How much did Maya save?

_____

Geometry

Circle the congruent figures.

1.

2.

3.

Algebra

Write >, <, or = to make each sentence true.

4. $10 + 10 \bigcirc 7 + 10$

5. $20 - 6 \bigcirc 30 - 16$

6. $5 + 16 \bigcirc 4 + 16$

7. $25 - 12 \bigcirc 25 - 4$

8. $50 - 25 \bigcirc 60 - 25$

9. $25 + 25 \bigcirc 20 + 30$

10. $\frac{1}{2} \bigcirc \frac{1}{4}$

11. $\frac{1}{4} \bigcirc \frac{1}{3}$

● Number and Operations

What part is gray? Write the fraction.

1.

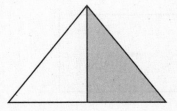

$\dfrac{1}{2}$ _____

2.

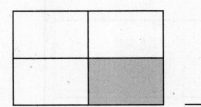

_____

3.

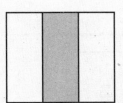

_____

4.

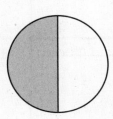

_____

Problem Solving

5. Frank is putting new tile in his bathroom. He will need 5 rows of tiles. Each row will have 6 tiles in it. How many tiles does Frank need in all?   _____ tiles

6. Sylvia arranged photos on 8 pages of her photo album. She put 4 pictures on each page. How many photos did Sylvia put in her album?   _____ photos

7. Carmen did her homework in 40 minutes. Bob did his homework in half that time. How long did it take Bob to do his homework?   _____ minutes

Name _____ Date _____

Measurement

What is the area of each shaded figure?
Each ☐ is 1 square unit.

1.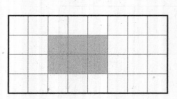

_____ square units

2.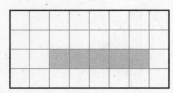

_____ square units

3.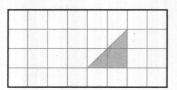

_____ square units

4.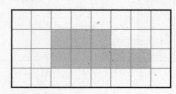

_____ square units

Number and Operations

What number is missing?

5.

$90 + \underline{\hspace{1cm}} = 100$

6.

$25 + \underline{\hspace{1cm}} = 100$

7.

$\underline{\hspace{1cm}} + 50 = 100$

8.

$\underline{\hspace{1cm}} + 31 = 100$

Measurement

1. What time is it? Follow the arrows.

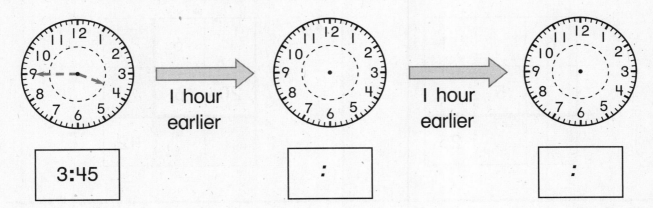

| 3:45 | : | : |

Number and Operations

2. Add. What are the missing amounts?

+	Q	D	N	P
Q	50¢			
D				
N				
P				

Number and Operations

What numbers are missing?

1.

6		11
	5	
		25

2.

	6	
20	6	
		42

3.

	25	35
20		
	55	85

4.

	7	
	5	45
80		92

Problem Solving

Write a number sentence to solve each problem.

5. Art has 35 balloons. He gives some balloons away. He has 16 balloons left. How many balloons does Art give away?

_____

_____ balloons

6. Mike drives 75 miles from home to the lake. Then he drives another 23 miles to the cabin. How many miles does Mike drive in all?

_____

_____ miles

● Number and Operations

Complete each fact family.

1. $9 + \underline{\hspace{2cm}} = 16$

   $\underline{\hspace{2cm}} + 9 = 16$

   $16 - 7 = \underline{\hspace{2cm}}$

   $16 - \underline{\hspace{2cm}} = 7$

2. $8 + 5 = \underline{\hspace{2cm}}$

   $5 + \underline{\hspace{2cm}} = 13$

   $13 - 8 = \underline{\hspace{2cm}}$

   $13 - \underline{\hspace{2cm}} = 8$

3. $14 - \underline{\hspace{2cm}} = 9$

   $14 - 9 = \underline{\hspace{2cm}}$

   $\underline{\hspace{2cm}} + 5 = 14$

   $5 + 9 = \underline{\hspace{2cm}}$

4. $\underline{\hspace{2cm}} - 7 = 5$

   $12 - 5 = \underline{\hspace{2cm}}$

   $7 + \underline{\hspace{2cm}} = 12$

   $5 + 7 = \underline{\hspace{2cm}}$

Geometry

Solve.

5. What new figures do you get if you cut on the dashed lines?

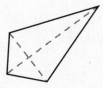

_____

6. Jim cut this figure into two equal pieces. He cut one straight line. How many sides does each piece have?

_____

Data Analysis and Probability

I. This table shows how many muffins were sold at a bake sale in one week. Make a line graph of the data.

Muffins Sold at the Bake Sale	
Day	Number of Muffins Sold
Monday	7
Tuesday	9
Wednesday	3
Thursday	8
Friday	10

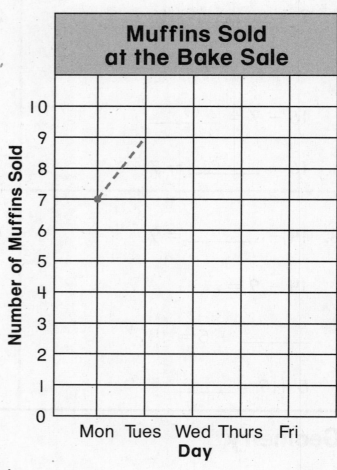

2. How many more muffins were sold on Thursday than on Wednesday? _____ more muffins

3. How many muffins were sold during the week? _____ muffins

4. Use the graph to guess how many muffins might be sold on Saturday. Explain.

_____

_____

● **Algebra**

What is missing?

1. Rule: Add 4

2	3	4		6
6			9	

2. Rule: Subtract 3

7	9	11	13	
4				12

3. Rule: Add 2

2	3	4		6
4			7	

4. Rule: Subtract 5

8	11	14	17	
3				15

5. Rule: _____

3	4	5	6	7
6	8	10		

6. Rule: _____

6	8	10	12	14
3	4	5		

Algebra

What is missing? Make each sentence true.

1. $6 + 6 = 10 + $ _____

2. $15 + 4 < 15 + $ _____

3. $11 - 3 > $ _____ $- 3$

4. $18 - 12 = $ _____ $+ 3$

5. $2 \times 7 < $ _____ $\times 7$

6. $5 + 3 > 5 + $ _____

7. $4 - 4 < 4 - $ _____

8. $8 - 3 = $ _____ $- 2$

9. $9 + 5 > 10 + $ _____

10. $12 + 6 = $ _____ $\times$ _____

Number and Operations

What part of the picture is gray?
Complete each fraction.

11.

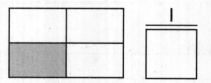

12.

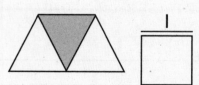

13.

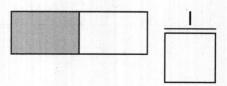

14.

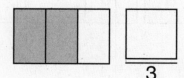

● Number and Operations

Name the fewest bills and coins that show the amount.

I. $3.27

_____

_____

2. $1.23

_____

_____

3. 68¢

_____

_____

4. $2.55

_____

_____

What is missing? Complete each puzzle.

5.

40		
	7	
60		

6.

60	13	73
	8	38

Measurement

How long is the picture of each object?
Use a ruler to measure to the nearest inch.

1.

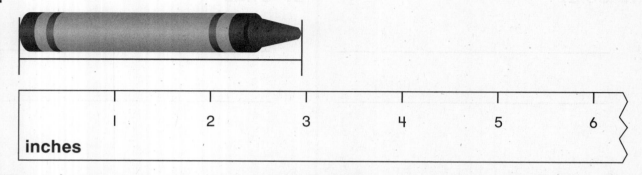

| inches | 1 | 2 | 3 | 4 | 5 | 6 |

2.

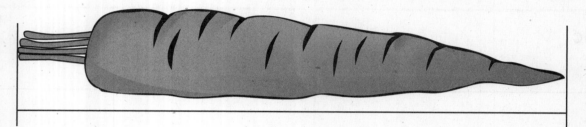

about _____ inches

Number and Operations

How many in all?

3.

_____ counters

4.

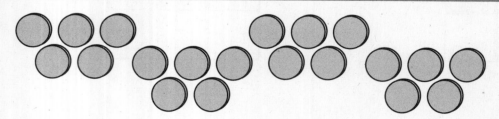

_____ counters

● Algebra

What is missing from each table? Use the rule.

1.

a	5	10	14	19	22		30	34	
$a + 4$	9		18			29			40

2.

a	20	32	45	49		66		85	
$a - 8$		24			50		62		84

Measurement

Circle the better estimate for each time.

3. About how long does it take to eat a peach?

    I second   I minute

4. About how long might it take to paint a large picture?

    I minute   I hour

Data Analysis and Probability

There are 10 cubes in each bag. Write how many and color the cubes to match the story.

1. Sue is **more likely** to pull a red cube from the bag.

   _____ red   _3_ green

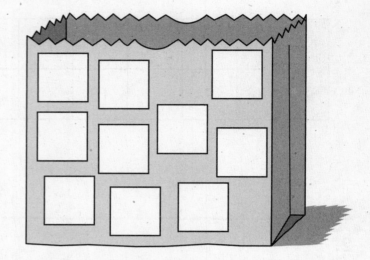

2. Stan is **equally likely** to pull an orange or black cube from the bag.

   _____ orange   _____ black

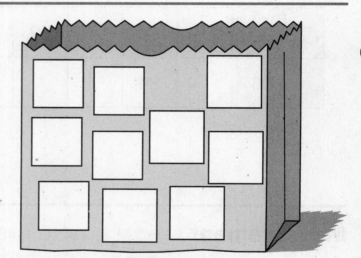

Algebra

3. Write the tens that sandwich 34. Which is closer?

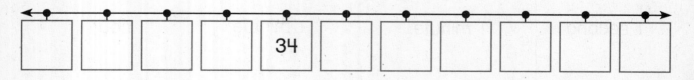

The nearest ten to 34 is _____.

● Algebra

What comes next? Continue each pattern.

1.

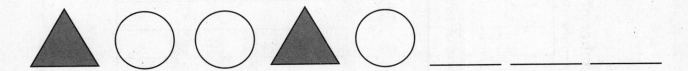

___ ___ ___ ___

_____

2. 5, 10, 15, 20, 25, 30, _____, _____, _____, _____

_____

3.

12:15	1:15	2:15			

_____

Problem Solving

4. Sebastian has 3 quarters and a nickel. Mick has
6 dimes and 3 nickels. How much more money
does Sebastian have than Mick?   _____¢

5. A classroom has 5 rows of desks with 6 desks
in each row. How many desks are there in all?   _____ desks

Data Analysis and Probability

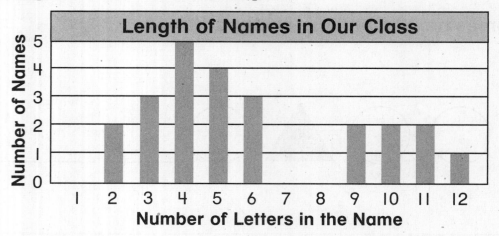

Length of Names in Our Class

Number of Names (y-axis): 0, 1, 2, 3, 4, 5

Number of Letters in the Name (x-axis): 1, 2, 3, 4, 5, 6, 7, 8, 9, 10, 11, 12

1. How many letters are in the longest name? _____ letters

2. How many names have 4 letters? _____ names

3. How many letters are in the names for the most children? _____ letters

Measurement

What time is it?

4.  → **5 minutes later**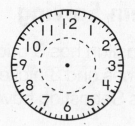

5.  → **5 minutes later**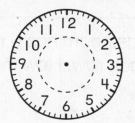

● Number and Operations

Add.

1.

31	26	44	32	15
42	41	30	42	60
+ 15	+ 20	+ 25	+ 22	+ 12

2.

16	21	24	10	41
30	13	31	37	2
+ 33	+ 4	+ 42	+ 30	+ 50

3.

2	21	31	10	12
71	16	21	20	31
5	12	11	30	14
+ 10	+ 10	+ 1	+ 40	+ 41

Problem Solving

4. Cass has 25 tiles. She puts them into equal rows. How many tiles are in each row?

_____ tiles

5. George puts all of his tiles in 4 rows of 6. How many tiles does he have in all?

_____ tiles

Geometry

**Match each three-dimensional figure
to its faces.**

1.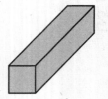

2.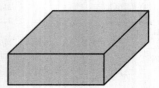

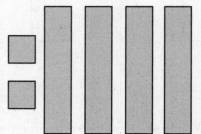

3.

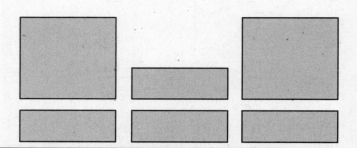

Problem Solving

4. Tina is building a fence all around
   her yard. The yard is square. Each
   side is 12 feet long. What will be
   the length of the whole fence?

   _____ feet

Name _____ Date _____

Number and Operations

How much money is there?

1.  _____

2.  _____

3.  _____

Algebra

Write the missing numbers.

4. $86 + 6 = 80 + \boxed{} = 90 + \boxed{} = \bigcirc$

5. $59 + 2 = 50 + \boxed{} = 60 + \boxed{} = \bigcirc$

6. $35 + 7 = \boxed{} + 12 = \boxed{} + 2 = \bigcirc$

7. $77 + 6 = \boxed{} + 13 = \boxed{} + 3 = \bigcirc$

Measurement

Measure each string to the nearest centimeter.

1.

about _____ centimeters

2.

about _____ centimeters

3.

about _____ centimeters

4.

about _____ centimeters

5.

about _____ centimeters

Number and Operations

What numbers are missing?

6.

	3	83
90		100

7.

20		29
	10	100

● Algebra

1. Follow the rule. Write each missing number.

m	19	27	43			38		
m + 10	29	37	53	16			44	
m + 8	27	35	51		85			20

Number and Operations

Find each difference.

2. $92 - 2 =$ ☐

3. $92 - 4 =$ ☐

4. $92 - 6 =$ ☐

5. $17 - 7 =$ ☐

6. $17 - 9 =$ ☐

7. $17 -$ ☐ $= 7$

8.
$$\begin{array}{r} 54 \\ -\ \ 4 \\ \hline \boxed{} \end{array}$$

9.
$$\begin{array}{r} 54 \\ -\ \ 6 \\ \hline \boxed{} \end{array}$$

10.
$$\begin{array}{r} 89 \\ -\ \ 7 \\ \hline \boxed{} \end{array}$$

11.
$$\begin{array}{r} 89 \\ -\ \ 8 \\ \hline \boxed{} \end{array}$$

12.
$$\begin{array}{r} 77 \\ -\ 10 \\ \hline \boxed{} \end{array}$$

13.
$$\begin{array}{r} 77 \\ -\ \ 9 \\ \hline \boxed{} \end{array}$$

Algebra

I. Follow the rule. Write each missing number.

a	100	10	30	60	20	50	90	70
$100 - a$	0	90	70					

Use the number line to add and subtract.

2.

0 2 4 6 8 10 12 14 16 18 20

☐ + 6 = 8   ☐ + 6 = 14   ☐ + 6 = 10

☐ − 4 = 8   ☐ − 4 = 14   ☐ − 4 = 10

3.

0 3 6 9 12 15 18 21 24 27 30

☐ + 6 = 30   ☐ + 9 = 15   ☐ + 12 = 21

☐ − 6 = 18   ☐ − 9 = 0   ☐ − 15 = 12

● **Geometry**

How many faces does each figure have?

	Figure	Number of Faces
1.		
2.		
3.		
4.		

Algebra

Fill in the missing numbers for the jumps.

5.

0   4   ____   ____   ____

6.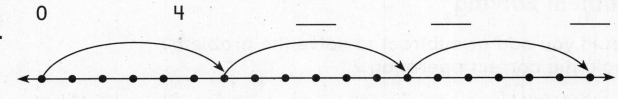

6   12

7.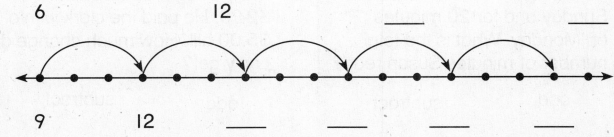

9   12   ____   ____   ____

Data Analysis and Probability

1. Use the tally table to complete the pictograph.

Apples Eaten by Mrs. Liu's Class	
Day	**Tally**
Monday	卌 卌 ‖
Tuesday	卌 卌 卌 卌
Wednesday	卌 ‖‖

Apples Eaten by Mrs. Liu's Class										
Monday	🍎									
Tuesday										
Wednesday										

Key: Each 🍎 stands for 2 apples.

Problem Solving

**Would you add or subtract to solve the problem?
Circle the correct operation.**

2. Susan read for 30 minutes on Sunday and for 20 minutes on Monday. What is the total number of minutes Susan read?

add   subtract

3. Larry bought a notebook for $2.95. He paid the clerk with a $5.00 bill. How much change did Larry get?

add   subtract

● Number and Operations

Add or subtract.

1. $\begin{array}{r} 118 \\ +262 \\ \hline \end{array}$	2. $\begin{array}{r} 531 \\ +208 \\ \hline \end{array}$	3. $\begin{array}{r} 356 \\ +325 \\ \hline \end{array}$	4. $\begin{array}{r} 68 \\ +141 \\ \hline \end{array}$
5. $\begin{array}{r} 890 \\ -371 \\ \hline \end{array}$	6. $\begin{array}{r} 785 \\ -371 \\ \hline \end{array}$	7. $\begin{array}{r} 478 \\ -371 \\ \hline \end{array}$	8. $\begin{array}{r} 602 \\ -371 \\ \hline \end{array}$

● Measurement

Which is the best unit? Write *inches*, *feet*, or *yards*.

9.

The desk is about 3 _____ wide.

10.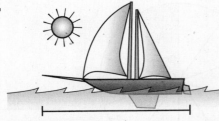

The boat is about 5 _____ long.

11.

The paintbrush is about 8 _____ long.

Name _____ Date _____

Measurement

What time is it?

1.

_____ minutes after _____

2.

_____ minutes after _____

3.

_____ minutes before _____

4.

_____ minutes before _____

Number and Operations

Write the fraction for the gray part.

Circle if it is closer to 0, $\frac{1}{2}$ or 1.

5.

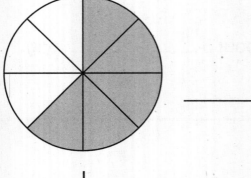

_____

0      $\frac{1}{2}$ 1

6.

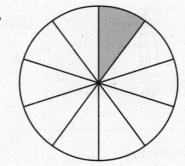

_____

0      $\frac{1}{2}$ 1

© Education Development Center, Inc.